New Rochelle Public Library

SEP __ '05

W9-BRU-968

Nature's Children

JELLYFISH

James Kinchen

GROLIER
EDUCATIONAL

FACTS IN BRIEF

Classification of Jellyfish

Phylum: *Coelenterata*
Class: *Schyphozoa* (true jellyfish)
Order: There are five orders of jellyfish.
Species: There are about 200 species of jellyfish.

World distribution.   Seas and oceans all over the world. The largest species inhabit cold, Arctic waters.

Habitat.   Open ocean, either near the surface or in deep water. A few species live close to shore.

Distinctive physical characteristics.   Delicate, gel-filled body; umbrella shaped, with a mouth at the center; outer surfaces covered with sting cells; often have long, trailing tentacles.

Habits.   Swim weakly by rhythmically contracting their bodies. Harpoon prey with sting cells. Carried by ocean currents and often washed ashore in large numbers after storms. Reproduce by laying eggs. Complex life cycles involving up to four stages.

Diet.   Plankton, small fish, and crustaceans.

All rights reserved. Except for use in a review, no part of this book may be reproduced, stored in a retrieval system, or transmitted in any form, or by any means, electronic, mechanical photocopying, recording, or otherwise, without prior permission of the publisher.

© 1999 Brown Partworks Limited
Printed and bound in U.S.A.
Editor: James Kinchen
Designer: Tim Brown

Published by:

GROLIER
EDUCATIONAL

Sherman Turnpike, Danbury, Connecticut 06816

Library of Congress Cataloging-in-Publishing Data
Jellyfish.
    p. cm. -- (Nature's children. Set 6)
    ISBN 0-7172-9360-2 (alk. paper) -- ISBN 0-7172-9351-3 (set)
    1. Jellyfish--Juvenile Literature. [1. Jellyfish.] I. Grolier Educational (Firm) II. Series.
QL377.S4K56 1999
593.5'3—dc21
    98-33395

Contents

It is evening on the coral reef. Small fish, eager to find a safe place to pass the night, dart in and out of the coral crevices. Above them a barely visible shape pulses through the clear water. It has no eyes, no brain, and its delicate body is made almost entirely of water. It seems impossible that this fragile animal— a jellyfish—could even survive among the crashing waves, yet it is one of the most deadly of the ocean's hunters.

Jellyfish are simple animals that were swimming in the oceans millions of years before the dinosaurs. They have changed little since then because they are so well adapted to their way of life. Jellyfish and their relatives have colonized almost all of the water on Earth. Some are so small that they cannot be seen without a magnifying glass, while others build coral reefs that are visible from space. Read on to find out more about this amazing family of animals.

A jellyfish drifts past brightly colored fish and corals.

The Jellyfish Family

Jellyfish belong to a large group of animals called Coelenterata. You can see many of the members of this family if you visit the seashore. The flowerlike sea anemones that grow in rock pools, the corals that inhabit warm, tropical seas, and the delicate, plantlike sea firs that grow on rocks and seaweed are all members of this group of animals.

Although jellyfish are often found washed up on beaches or trapped in rock pools, they are most at home in the open ocean, drifting just below the surface. Most jellyfish range from half an inch (12 millimeters) to one foot (30 centimeters) across. The largest type, the cyanea, or sea blubber, jellyfish, can grow to six feet (two meters) wide—the size of a small boat. The tentacles that trail behind these monsters can be 117 feet (34 meters) long!

Corals like these may look like colorful rocks, but in fact they are colonies of tiny jellyfish relatives.

Delicate Bodies

The jellyfish family name, Coelenterata, comes from a Greek word meaning "hollow gut." True to their name, jellyfish have hollow, baglike bodies with a single opening through which they eat and dispose of waste.

Inside our bodies there are many different organs, each with its own function. Lungs absorb oxygen from the air, the heart pumps blood around the body, the stomach helps us digest our food, and our brains are responsible for thinking and controlling all the other parts. A jellyfish's body, however, is made up of just two skinlike layers with a space filled with gel in between. The outer skin contains stings for defense and sense cells that can detect light and gravity, giving the jellyfish its sense of direction. The inner skin, similar to our stomachs, is responsible for absorbing food. Between the two skins runs a network of nerves and muscles that, together, can open and close the jellyfish's body like an umbrella, pushing it gently through the water.

Opposite page: *Jellyfish fossils like this one are rare because the delicate bodies of jellyfish do not often get preserved in rocks.*

9

Fantastic Shapes

Opposite page: *Washed up on a beach, a beautiful jellyfish will melt away in a few hours, leaving only a sticky puddle.*

Jellyfish live in seas and oceans all over the world, from the icy waters of the Arctic and Antarctic Oceans to warm, tropical seas. Most of the animal life in the oceans lives near the surface, but a few types of jellyfish can be found in the ocean depths, more than 3,300 feet (1,000 meters) below the surface.

When you see a jellyfish washed up on a beach, it looks like a shapeless, watery mass. However, in their natural environment—the ocean—jellyfish are supported by the sea water and can adopt a wide variety of shapes to suit their different ways of life. Those that feed on plankton—a mixture of tiny plants and animals living near the ocean's surface—waft water past their mouths and strain out the tiny creatures with their short, sticky tentacles. Other types that feed mainly on larger prey, such as fish, have longer tentacles covered with powerful stings.

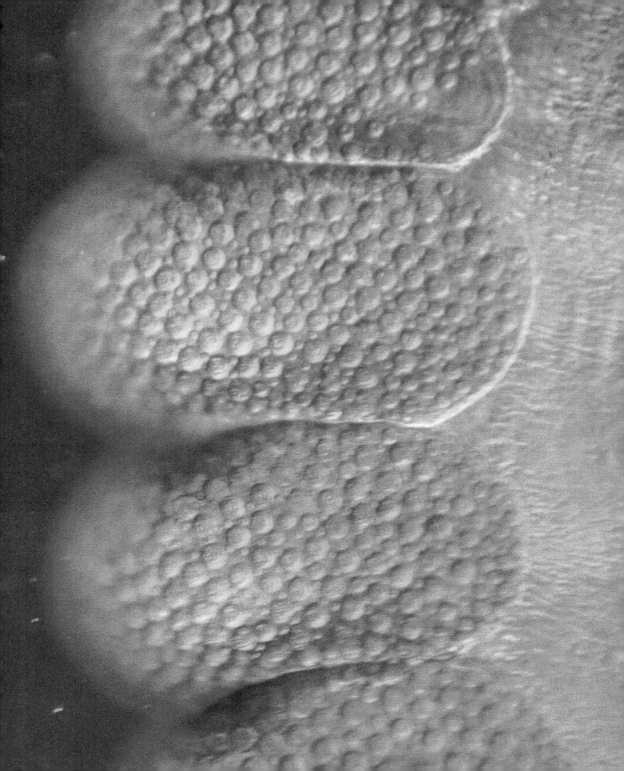

Poisonous Stings

The first thing we learn about jellyfish is to be wary of their stings. Many of the large types have powerful stings that can cause skin rashes or more severe injuries. A sting from a sea wasp, a type of jellyfish common around the coast of northern Australia, can kill a human in less than 15 minutes.

Jellyfish use their stings to capture prey and to defend themselves against enemies. A jellyfish sting looks like a miniature, hollow harpoon. The harpoon is kept tightly coiled up inside a sting cell, called a nematocyst. The nematocyst is triggered by touch. As soon as the jellyfish brushes against something, prey or enemy, its nematocysts fire their harpoons. Each harpoon uncoils to 50 times its original length, and vicious barbs ensure that it gets a firm hold. Poison starts to flow down the harpoon into the unfortunate victim, causing it extreme pain. Large animals may be able to struggle free, but for small fish or shrimps it may already be too late.

Opposite page:
The sting cells of a Portuguese man-of-war. Inside each there is a tightly coiled harpoon, ready to fire.

Trailing Tentacles

Medusa was a beautiful but deadly monster from an ancient Greek myth, or story. She had the body of a woman, with a writhing mass of snakes for hair. A single look from her was enough to turn a person into stone. In some ways jellyfish resemble this make-believe creature. They, too, have beautiful bodies, and their trailing tentacles can deliver a painful sting. In fact, scientists call adult jellyfish "medusas." Do you think the name suits them?

Jellyfish use their tentacles to catch the animals on which they feed. Usually, there are four or eight large tentacles arranged in a circle around the jellyfish's mouth, in the center of its umbrella-shaped body. Most types of jellyfish also have a ring of shorter, thread-like tentacles around the edges of their bodies.

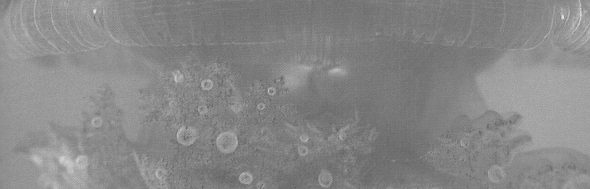

This jellyfish, living near the Great Barrier Reef off the coast of Australia, has thick, frilly tentacles that it uses to catch its food.

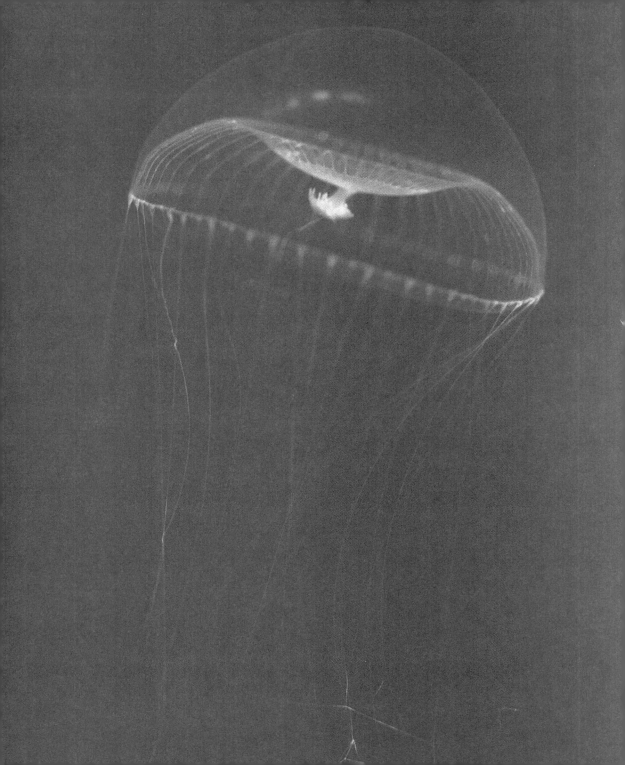

Unseen Danger

Have you ever wondered why soldiers wear drab, brown, or green clothing, covered in irregular patterns? These colors, called camouflage, help them blend into their surroundings, making them more difficult for the enemy to detect. Animals use camouflage, too. A leopard's spotted coat makes it almost invisible against the dappled sunlight of the forest. A stick insect, motionless in a bush, is easily missed by passing predators. And the see-through body of a jellyfish is perfect for hiding in the clear water of the ocean.

Jellyfish are not fast swimmers, so their only way of avoiding enemies and catching prey is to be invisible. Some types do have bright colors in small patches on their bodies or at the edges of their tentacles. Rather than revealing the jellyfish, these patches can actually help it catch prey. Mistaking the patch for a small piece of food, a fish may stray too close and become trapped in the jellyfish's tentacles.

*Opposite page:
A crystal jellyfish is almost invisible in the clear water of Monterey Bay in California.*

Deadly Hunters

Each jellyfish has its own method of catching other animals for food. The moon jelly feeds on tiny animals in the plankton. Its short tentacles are covered in minute hairs called cilia. By beating these hairs very fast, it wafts a current of water past its mouth. Food particles become trapped in the sticky mucus that covers the jellyfish's body. The cabbage blebs jellyfish has several thousand small mouths dotted over its tentacles. The mouths act like a sieve, filtering plankton from the water.

Jellyfish that catch larger prey, such as fish and shrimp, have long tentacles that trail behind them. The tentacles are difficult to see underwater, and the prey often swims into them by mistake. The sticky tentacles are covered in stings, which quickly overpower the jellyfish's victim.

Swimming Slowly

Jellyfish are not strong swimmers, but they are determined! They can pulse slowly through the water for hours or even days without taking a break. A jellyfish swims by rhythmically opening and closing its body like an umbrella. Its network of muscles pulls the umbrella closed, then the gel between its skin makes the umbrella spring back open.

If you want to see how a jellyfish moves, try filling a balloon with water. Using your hands, you can squeeze the balloon into a new shape, which is what the jellyfish's muscles do. When you let go, the balloon will quickly return to its original shape.

While jellyfish never chase their prey, swimming helps them stay in areas that are rich in food, gather together in shoals for breeding, and stop being washed ashore.

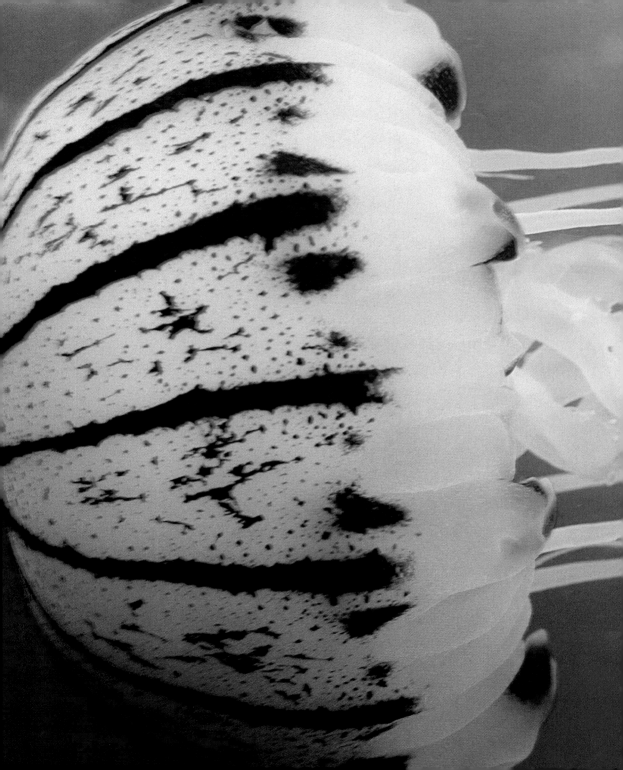

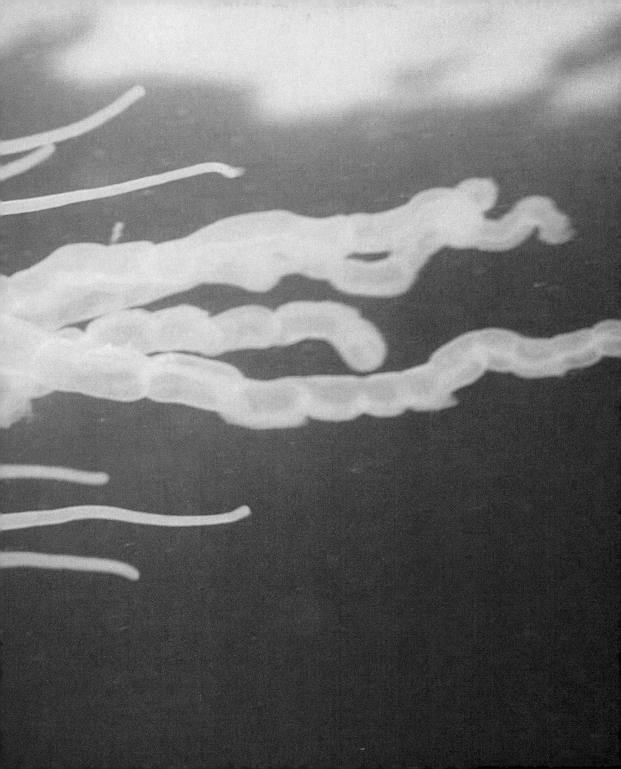

Swept Along

Opposite page: *Occasionally storms or strong currents carry swarms of jellyfish toward the shore, making the water unsafe for human bathers.*

The water in the oceans does not stay in one place. Driven by the wind, mighty currents flow like giant rivers around all of the world's oceans. The Humbolt Current in the Pacific Ocean carries cold water from the Arctic down the West Coast of the United States. The Gulf Stream in the Atlantic Ocean brings warm water from the tropics to the coast of Britain and northern Europe. These currents are vital for life in the oceans. They carry nutrients and oxygen between the seabed and the surface. They can even change the weather! The El Niño effect, which causes floods and droughts in many countries, occurs when the current in the Pacific Ocean changes direction.

Jellyfish, and many other animals, are swept along by ocean currents. They rely on the currents to carry them to areas in which they can breed or find food. It can be risky— sometimes thousands of jellyfish are swept ashore after storms. Others are carried to barren areas of the ocean where they starve.

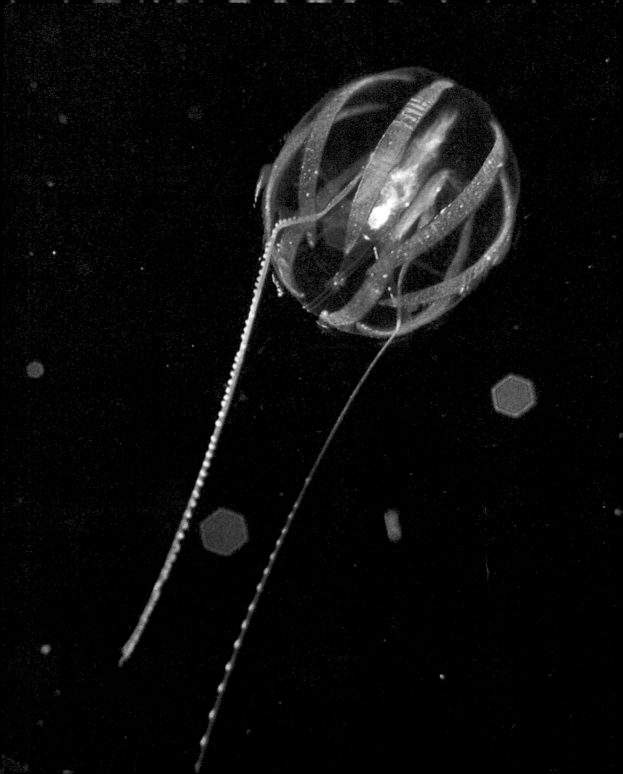

Glow in the Dark

Many types of jellyfish are able to make their own light. Sometimes, on calm nights, the Atlantic Ocean is dotted with pale green, glowing pelagia jellyfish that gather near the surface in large shoals. Although no one really knows why these jellyfish glow, it may be a way of attracting prey. Some of the small creatures in the plankton are attracted toward bright lights. During the day they swim upward toward the bright surface, where there is plentiful food and oxygen. At night, however, they may mistake a bright shoal of jellyfish for the surface and swim into their tentacles.

Close relatives of the jellyfish, the comb jellies, can emit a bright flash of light if they are disturbed. In tropical seas, where these animals are common, divers often leave trails of glowing comb jellies behind them as they swim in the water.

Startled by a passing diver, a comb jelly, or sea gooseberry, glows like a miniature lightbulb.

Hangers On

While most sea creatures do their best to avoid jellyfish, there are a few animals that like to come in contact with them. Several types of small fish are often seen swimming safely beneath large jellyfish. They even rub their bodies against the stinging tentacles without seeming to come to any harm. Jellyfish are covered with slimy mucus that stops them from accidentally stinging themselves. By rubbing this mucus over their own bodies, the fish prevent the jellyfish's stings from firing.

Would you want a slimy, dangerous jellyfish for a neighbor? For a fish there are a number of benefits. Hiding among the jellyfish's tentacles, it is safe from attack by predators, and there are always tasty tidbits left over from the jellyfish's meals.

A pair of shepherd fish shelters among the tentacles of a Portuguese man-of-war.

Jellyfish Enemies

Jellyfish are eaten by a few types of animals that are not affected by their stings. Sea turtles have armored, bony jaws that allow them to feast on even the most poisonous of jellyfish. In fact, the largest of all sea turtles, the giant leatherback turtle, eats almost nothing else. Crabs, too, are often seen digging their claws into the soft flesh of jellyfish that have been washed into shallow water. Attacks from crabs rarely kill jellyfish, because injuries to their simple bodies heal quickly.

Many sea slugs also feed on jellyfish and their relatives, the sea anemones. A sea slug has no armor for protection, but its body is covered in a special mucus that prevents the jellyfish's stings from firing. Instead of digesting the jellyfish's stings, the sea slug steals them for its own protection. The sting cells are stored on its back, where they will sting any fish that tries to eat the sea slug.

A sea slug feasts on a by-the-wind-sailor jellyfish.

Man-of-War

Opposite page:
Stay back! The coiled tentacles of the Portuguese man-of-war can deliver a lethal sting.

One of the most feared of all jellyfish, the Portuguese man-of-war, is not a true jellyfish at all! If you look closely, you will see that instead of an umbrella-shaped body, it has a float filled with gas. True jellyfish swim through the water, while the Portuguese man-of-war is blown across the surface by the wind, like a sailing boat. A Portuguese man-of-war is not even a single animal but is a colony of tiny individuals, called polyps, each with its own job to do. Some polyps keep the float in good condition and full of gas, while others catch food for the whole colony.

The Portuguese man-of-war deserves its fearsome reputation. Its tentacles, which can grow to 45 feet (14 meters) long, are towed behind it, like a fishing boat towing a net. It eats small fish and shrimp, which it kills with its powerful stings. Mild stings from a Portuguese man-of-war cause a throbbing pain that can last for days. More serious stings can cause nausea or even death.

A sunny patch of seabed makes a perfect resting place for an upside-down jellyfish.

Upside Down

One of the most unusual members of the jellyfish family is the cassiopeia, or upside-down jellyfish. Unlike most jellyfish, it is usually found in shallow, tropical lagoons, close to shore. Its body is flattened, with a dense mass of short, green tentacles. As their name suggests, upside-down jellyfish spend most of their lives resting on their backs. Lying together on the bottom, they often look like beds of gently swaying seaweed.

The reason for this strange behavior lies in the way that cassiopeia feeds. Its tentacles, as well as catching small animals, house tiny plantlike organisms called algae. Like plants, algae can make the food they need to survive using sunlight. In exchange for their safe home, the algae share some of this food with the jellyfish.

Like a farmer, cassiopeia makes sure its algae are well looked after. It rests in places where there is plenty of light and continually wafts a stream of water over its little garden.

Laying Eggs

Opposite page:
The pale rings inside the bodies of female moon jellies produce eggs, while those of the males produce sperm.

Did you know that jellyfish can be male or female? Although it's not easy to tell the sexes apart, male jellyfish produce sperm, while females produce eggs. When they are ready to breed, most jellyfish gather together in large shoals. Then males release their sperm into the water, where it floats until it is absorbed by a passing female. Once it is inside the female's body, the sperm fertilizes her eggs.

After a few days developing inside the female's body or in special pouches in her tentacles, the eggs hatch. Each tiny baby jellyfish, called a larva, then floats out of its mother's mouth and swims away into the ocean. Jellyfish parents take no care of their young, and many babies will be eaten by predators or swept away on ocean currents. However, each female jellyfish can produce thousands of eggs. With so many babies, it is likely that many will survive to adulthood.

Taking a Break

The jellyfish larva does not look anything like its parents. It has no mouth, and its tiny, round body is propelled through the water by rows of hairlike cilia. Unable to feed, it drifts for several hours or even days before sinking to the seabed. Here, for the first time in its life, the jellyfish stops swimming and anchors itself firmly to a rock or piece of seaweed.

Once it has a firm hold on its new home, the baby jellyfish begins to change. The bottom part of its body grows into a long, slender stalk, while the top opens out into a ring of tentacles with a mouth at the center. At this stage the baby jellyfish is called a polyp. The jellyfish polyp is a successful hunter and grows quickly on a rich diet of tiny animals carried onto its stinging tentacles by the currents. Throughout this time it is building up reserves of food that will see it through the final and most dramatic change of its life.

The jellyfish polyp looks more like a sea anemone than an adult jellyfish.

Swimming Again

Some types of jellyfish spend only a few weeks as a polyp, while others may be anchored to the bottom for months. Eventually, however, the polyp undergoes another amazing change. First, the top part of its body divides so that it looks like a stack of dishes. Then, one by one, the dishes break off from the stalk and swim away. Each tiny dish is a new medusa, or adult jellyfish, so a single polyp can produce many new jellyfish.

The new medusas look more like snow-flakes than adult jellyfish. They swim back to the surface and join the plankton, where they feed on other small creatures, such as baby fish and shrimp. As they grow, their bodies fill out and their tentacles grow longer. It may be months or even years before they are old enough to lay eggs of their own.

A young jellyfish medusa drifts among lots of other small plankton creatures.

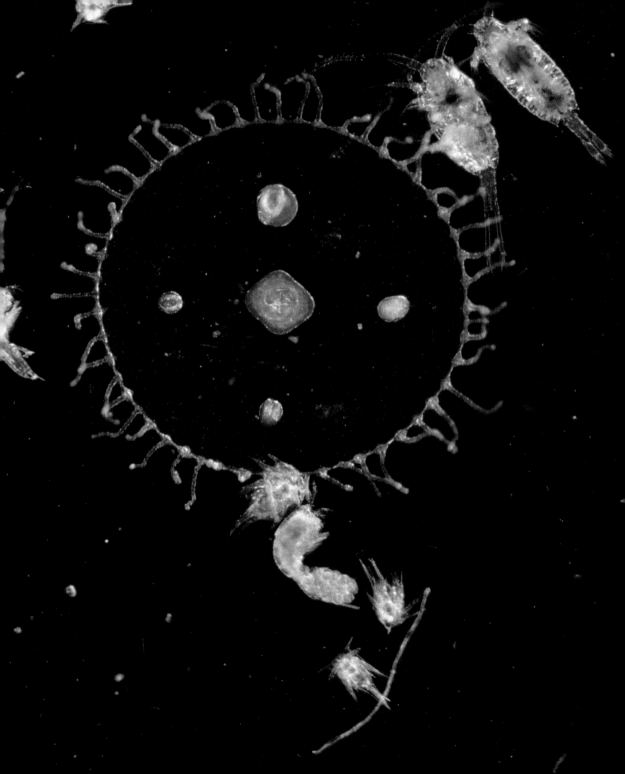

Life Cycles

The series of changes that an animal undergoes during its life is called a life cycle. The jellyfish life cycle consists of four different stages: egg, larva, polyp, and medusa. Our life cycle is far less complex. A baby human looks similar to an adult, only smaller. So why do jellyfish live in this complicated way? The answer is that their way of life enables them to produce a lot of new jellyfish very quickly. A single female can lay thousands of eggs, many of which will develop into polyps. A single polyp can divide into many new medusas. When conditions are favorable, jellyfish can grow and breed very rapidly.

Other members of the jellyfish family have similar life cycles. Sea anemones go through three stages: egg, larva, and adult. An adult sea anemone looks similar to a jellyfish polyp.

Like a jellyfish, this sea anemone started life as a swimming larva.

Reef Builders

Opposite page:
If you look closely at this piece of coral, you will see that it is made up of thousands of colorful polyps.

Have you ever snorkeled over a coral reef or seen one on television? Corals may look like colorful rocks, but in fact they are colonies of tiny jellyfish relatives. The individual animals that make up a coral, called coral polyps, look like tiny sea anemones. Each polyp surrounds its body in a hard casing that gives the coral its rocklike appearance. The polyps usually stay safely hidden away inside their homes during the day, but at night they extend their tentacles into the water to capture tiny food particles. Coral polyps also contain algae that can make food from sunlight.

Corals flourish in water that is clear, shallow, and warm. They need plenty of light for their algae, so they cannot live in deep or murky water. The largest coral reef in the world, the Great Barrier Reef off the east coast of Australia, is 1,200 miles (1,900 kilometers) long and can be seen from space.

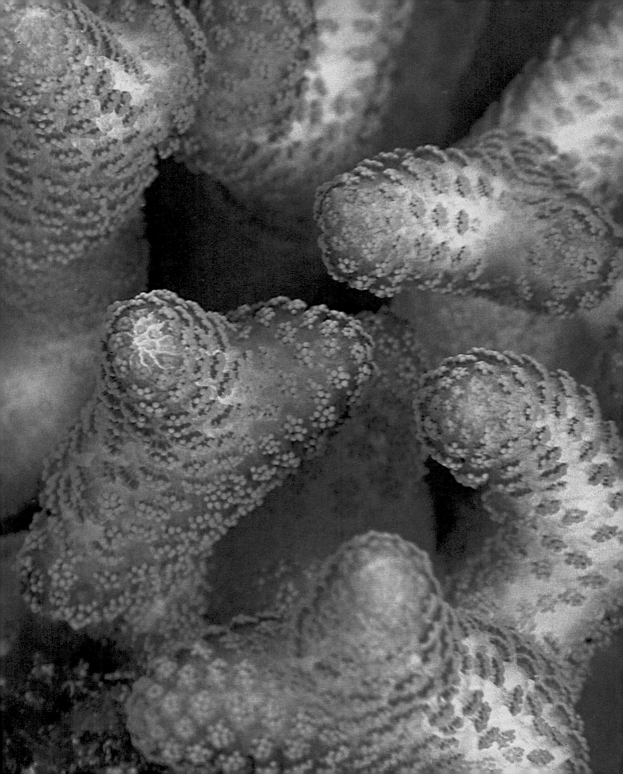

In the Pond

While almost all jellyfish live in the sea, there is one jellyfish relative that is found in freshwater ponds and streams. This unusual animal, called a hydra, looks like a jellyfish polyp. Hydras anchor their bodies to rocks and pondweeds. They feed by catching tiny creatures with their tentacles and by harvesting food from the green algae inside their bodies.

If you look at a hydra under a magnifying glass, you might see it doing somersaults! This is how the hydra moves around. First, it bends over and holds onto the surface with its tentacles. Then, it detaches its stalk and flips over to a new location.

Hydras can breed in two different ways. They can produce eggs and sperm like jellyfish, and they can produce young by budding. In this process a small lump appears on the stalk of the parent hydra. After a few days the lump grows tentacles, detaches from its parent, and starts to fend for itself.

Opposite page: *This adult hydra, which is about the length of your fingernail, has produced a bud that is almost ready to break away and start life on its own.*

Jellyfish in Danger

Jellyfish are remarkable creatures that have flourished for millions of years. They have outlived the dinosaurs and survived countless environmental changes. Today many of the threats that they face come from the activities of people. The delicate balance of the oceans is being disturbed by pollution and global warming. Changes in the temperature of the oceans can influence the currents that flow around them, with disastrous consequences for some species of jellyfish. Coral cannot survive in murky, polluted water, and many coral reefs are disappearing. As the reefs die, so do the fish and other animals that depend on the reefs for food and shelter.

The dumping of rubbish, such as plastic bags, into the sea is also harmful. Floating in the water, a plastic bag looks similar to a jellyfish. Sea turtles often eat plastic bags by accident and choke to death.

Words to Know

Algae   Simple plantlike creatures that use sunlight to make their food.

Camouflage   Natural disguise. Animals use camouflage to avoid being seen by predators and prey.

Coral reef   A large mass of corals living close together. Coral reefs form in warm, shallow, tropical waters where there is plenty of light.

Fossils   The remains of ancient creatures preserved in rocks.

Global warming   The gradual warming of the Earth due to harmful gases being released into the air.

Medusa   A mythical monster with snakes for hair. Scientists call adult jellyfish "medusas."

Plankton   A mixture of tiny animals and plants living close to the surface of the ocean.

Polyp   A young jellyfish at the stage when it is anchored to the seabed. Sea anemones and individual coral animals are also called polyps.

Predator   An animal that hunts other animals for food.

Prey   An animal that is hunted by other animals.

INDEX

Cover Photo: A.N.T. Photo Library / NHPA
Photo Credits: Stephen Frink / Corbis, page 4; Trevor McDonald / NHPA, page 7; Kevin Schafer / NHPA, page 8; A.N.T. Photo Library / NHPA, page 11; Image Quest 3-D / NHPA, pages 12, 35; Peter Parks / NHPA, pages 15, 27, 28; Rich Kirchner / NHPA, page 16; Brandon D. Cole / Corbis, page 20; Roy Waller / NHPA, pages 23, 40; Norbert Wu / NHPA, pages 24, 32, 36; Anthony Bannister / NHPA, page 31; Douglas P. Wilson; FLPA / Corbis, page 39; B. Jones & M. Shimlock / NHPA, page 43; M. I. Walker / NHPA, page 44.